Author Agent: longislandwriter@verizon.net

Copyright © Tim Roosevelt
ISBN-13: 978-1726117852
ISBN-10: 1726117855

Planes That Changed History

LOCKHEED U-2

Tim Roosevelt

Spy planes are different from other kinds of warplanes. They aren't like fighter planes. Fighter planes carry guns. They do tight turns in dogfights. Their pilots become aces. Spy planes do none of those things. Spy planes fly secretly. They fly in the middle of the night. They fly for hours in a straight line. That might not seem very exciting.

But spy planes are very important. Their job is "reconnaissance." That's a big word. It basically means gathering information by observing things. That's why they're called spy planes. One of the most famous spy planes in history is the Lockheed U-2, pictured above. The U-2 first flew during a time known as the Cold War.

To understand the U-2, it helps to know something about the Cold War. The Cold War began after World War Two ended. World War Two was fought between 1939 and 1945. In the early part of the war, horses were used to pull cannons. By the end of the war, atomic bombs had been used.

Those bombs marked a new age of weapons. During World War Two, the United States and the Soviet Union fought on the same side. But after the war, things changed. Both nations built nuclear weapons like the missile above. Such missiles could wipe out an entire city. The United States and the Soviet Union were no longer on the same side. They had become rivals. And they both feared each other's nuclear weapons.

This fear about nuclear weapons was called the Cold War. The Cold War wasn't a war where people shot at each other with guns. It was a "war" of fear between powerful nations. Both nations wanted to know what the other was doing. The United States felt that the only way to learn about Soviet nuclear missiles and bombers was to fly over the country.

But the Soviet Union is vast. You can't just take any plane and fly over a country to take top secret photos. It would be shot down. You would need a special plane. And that's where the Lockheed U-2 entered the story. It could reach very high altitudes. Altitude means height above the earth.

This ability to fly high would allow it to enter the Soviet skies and take photos of missiles and bombers. The photo above shows what a U-2 pilot sees when the plane is high above the earth. The photo on the opposite page shows the U-2 from the outside. This plane is one of the modern U-2s. It is much larger than the early U-2 on page one.

The U-2 did not carry weapons to protect itself. It simply flew high. That was its main defense. Fighter planes carry many different weapons. They carry missiles. They carry cannons to use in dogfights. And such fighters even tried to shoot down U-2s. The U-2 was not built to shoot it out with fighter planes.

The U-2 was different from fast spy planes like the SR-71. That plane flew at over 2,000 mph. The photo above shows an SR-71. The U-2 flew at speeds that were not much faster than some propeller planes. But the U-2 was not built for speed. It was built to fly high. It was built to cover great distances.

Flying a spy plane is a tough job. That's true whether the spy plane is fast or slow. Pilots must plan their missions carefully. They worry about running out of gas. On a plane, they say "fuel" instead of gas. Pilots must sit for hours without moving around. They are strapped to their seats. They wear big, bulky outfits. They breathe oxygen through a helmet which is part of that big outfit. It isn't comfortable.

Flying a spy plane is also dangerous. Spy planes take top secret photos. They count enemy ships, planes, and troops. Countries don't like having spy planes in their skies. They try to shoot down spy planes. That's why it's dangerous work. But it's important work.

Spy planes take photos that look like the one above. Spy plane cameras are very powerful. In the old days, U-2 cameras would bring back photos on large rolls of film. After the U-2 landed, experts would study the film. They would look for important things. Where is that ship going? How many planes are on that runway? Are those trucks carrying missiles? What has changed since the last photo?

Like most military planes, the U-2 came about in a contest. It was a contest where the government said it needed a new spy plane. The government told airplane companies how high the plane should fly. It told them what kind of camera it should carry. The companies then built a plane. They hoped the government would choose their plane over the others.

But it doesn't end there. After a plane wins the contest, it "evolves." That means it gets better. It does more things. The U-2 above is a modern one. The U-2 on page one is an early one. The modern one has bigger wings. It carries more equipment. It gathers more information. It "evolved" from the early U-2.

Lockheed entered the U-2 in this contest. Other companies entered their planes too. The other planes were similar to the B-57 Canberra pictured above. They had large wings. They had two engines. They could fly high.

The government liked two-engine planes. The second engine was like a spare tire. If one engine stopped working, the other would still get the plane home. But think about this. For a spy plane, its only defense is flying high. If a two-engine plane loses an engine, it loses altitude. That means it can't stay high in the sky. It took the power of both engines to stay way up there. So if the two-engine plane lost an engine, it was still in big trouble. Enemy missiles would probably shoot it down after it lost altitude.

The U-2 had only one engine. But that wasn't really a bad thing. In fact, it was good. That meant it was lighter. And light planes fly higher than heavy planes. The U-2 was based on the Lockheed F-104 Starfighter, pictured above. The planes look similar. Of course, the U-2 had much bigger wings.

The Starfighter could go 1,500 mph. The U-2 flew slower than 500 mph. The Starfighter was made to shoot down enemy planes. It flew fast because those small wings didn't create lots of drag. Drag is like friction. When you stick your hand out of a moving car, you feel air rushing around it. Your hand is small. It doesn't slow the car down much. But your hand creates drag by making wind swirl around it.

Now if you stuck a very large object out of a moving car, that would slow the car down. That's why the F-104 has small wings. The small wings don't create lots of drag. Large wings would create lots of drag and slow the F-104 down.

But there's something else to think about here, not just drag. And that force is *lift*. Yes, large wings slow a plane down. But large wings also give a plane lift. Lift is the force that pushes a plane upwards. And that's how the U-2 won the contest. A plane like the F-104 used its jet engine to reach sizzling speeds with small wings. The U-2 used its jet engine to reach soaring heights with huge wings. In the photo above, U-2s are at high altitude above California.

After the U-2 won the spy plane contest, the government had lots of work for it to do. Let's go back to the subject of the Cold War. Let's go back in time to the 1950s.

Remember the Cold War meant the United States and the Soviet Union were afraid of each other's nuclear missiles and bombers. U.S. President Eisenhower was told that the Soviet Union had more bombers and missiles than the United States. The word "gap" was used to describe these differences. People spoke of a missile gap. They spoke of a bomber gap. There really was no gap. Still, all this talk of "gaps" made President Eisenhower nervous.

President Eisenhower ordered U-2s to fly over the Soviet Union and take photos of missiles and bombers. The U-2s carried powerful camera sets like the one shown here. Before the U-2, the United States placed cameras on large balloons to take photos. That didn't work very well. You couldn't control where a balloon went. It just went where the wind took it. So the information from these balloons was not useful.

The leader of the Soviet Union was angry about these balloon flights and U-2 flights. He was Premier Nikita Khrushchev. Instead of president, Soviet people call their leader "premier."

You could even say Premier Khrushchev was furious about the U-2 flights. But President Eisenhower continued them. On May 1, 1960, the most famous U-2 flight in history took place. Pilot Francis Gary Powers would fly across the Soviet Union, called "U.S.S.R." on the map here. He would take photos. With these photos, the United States would learn about the Soviet missile program.

Mr. Powers took off from Pakistan in a U-2 like the one pictured above. It was an early model U-2. Pakistan is marked on the map by a yellow star with a red outline. He would fly toward the north. At the end of the long flight, he would land in Norway. Norway is marked by a yellow star with a green outline.

It was a very risky mission. Soviet radar was doing a good job of tracking U-2 flights by 1960. Soviet missiles had improved.

Those missiles could now reach high altitudes. Soviet fighter planes were also on the lookout. Near the city of Sverdlovsk, Mr. Powers was shot down by a missile. He was captured. The Soviet Union held a public trial. They put him in prison for espionage. Espionage means spying. He was later returned to the United States in exchange for a Soviet agent.

The map above shows the path of the flight. The mission was very long. It was very dangerous. Mr. Powers was a skilled and brave pilot. But the choice of dates was not a good one. May 1st is a Soviet holiday. This meant there would be fewer planes flying. And that would make it easier to find a lone spy plane in the sky.

The May 1960 U-2 incident marked the end of flights over the Soviet Union. But the U-2 flew in many other places. It flew in the Middle East in 1956 during the Suez Crisis. It flew during the Cuban Missile Crisis in 1962. More recently, it flew in Iraq. It flew in Afghanistan.

The U-2 went through many changes over the years. It got bigger. It became more modern. It was given video displays. People on the ground didn't have to wait for U-2s to land with their large rolls of film. The plane could send information while it was still flying. One thing that didn't change was the wing supports. These are called pogos. In the photo above, a pogo is lifted into place. They support the U-2's wings until the plane takes off. You can see pogos under the wings of an early U-2 on page 13.

Landing a U-2 can be tricky. Those giant wings give the U-2 lots of lift. That's good for reaching high altitudes. But all that lift makes landing the U-2 difficult. It takes lots of concentration to make a gentle landing.

The strong lift created by the U-2's large wings could make the plane bounce up and down. Coming down hard could damage equipment. The problem is visibility. It's hard for U-2 pilots to get a good view of the outside. A chase car helps here. It follows the incoming U-2. The chase car tells the U-2 pilot how high above the runway the plane is. This helps make a smooth landing. Chase cars need to be fast to keep up with incoming U-2s. Yes, that's a Camaro, and it's fast!

Let's talk about the color of U-2s. You probably noticed that all the modern U-2s in this book are painted black. The early U-2s (smaller ones) came in different colors. Some were black. Some were dark blue. On page one, the U-2 has camouflage paint. That means using different color patterns to make a plane blend in with the background. Some U-2s were left in bare metal, like the one with the camera set on page 12. That keeps a plane light, since paint adds weight. And being light helps reach high altitudes.

Why are dark colors used on spy planes? Spy planes fly at high altitudes. They fly where the color of the sky becomes a deep blue. Look at the U-2 above. Against the dark sky, it blends in well. It's hard for a fighter pilot to spot it.

When does the meal cart come out on this plane? If you ever went on a long flight, you were probably served a meal. That makes a flight nicer. Well, U-2 pilots also have meals when they fly. But they're nothing as fancy as an airline meal.

The food comes as a paste in tubes. Pilots squeeze the tubes to eat. They suck the food through a straw that goes through a hole in their helmets. You can see the choices above. There is even a caffeinated apple pie. "Caffeinated" means it has caffeine, the thing in coffee that makes it perky. It's like having apple pie and coffee together. That sounds safer than having a hot cup of coffee in your lap at 70,000 feet!

Before a U-2 takes off, there's a list of things to be done. The pilot's helpers inspect the pressure suit. They check it for leaks. They help the pilot with the suit. It is bulky and difficult to put on. But the suit keeps a pilot alive at 70,000 feet.

Pilots get a physical exam. That's to make sure they're healthy enough to fly. Even for a fit person, it's hard to be strapped into a plane for five or six hours. If someone has a cold, bad toothache, or the flu, that could make things even harder. The pilot will start to breathe pure oxygen an hour before the flight. This is to get nitrogen out of the blood. That prevents the pilot from getting sick.

After the physical exam, the pilot is taken to the U-2. The helpers strap the pilot into the plane. They check oxygen hoses and other connections. The U-2 slowly makes its way onto the runway. The ground crew signals to the plane. The pilot pushes the throttle forward and the General Electric turbine spins faster and faster. The plane gains speed as it hurtles down the runway.

Soon the U-2 is off the ground with a thunderous roar as it climbs toward 70,000 feet. The world will seem peaceful at that altitude. The gently curved horizon will glow brightly against the deep blue darkness of space in the distance. And for a few hours, the U-2 and its pilot will be in this world.

Planes That Changed History - Lockheed U-2

THE END

The Lockheed U-2 has gone through many changes since it was first produced in the 1950s. Francis Gary Powers was shot down over the Soviet Union while flying a U-2A. That early model U-2 seems like a totally different airplane than the huge U-2Ss of today. The plane's fuselage (body) has grown along with its wingspan. It was given a GE F118-101 engine to replace the J57 and J75 engines found in earlier models. The largest changes are in the plane's powerful electronics. The U-2 now has the ability to send information to people on the ground while it is still in the air. With early U-2s, people on the ground would wait for the plane to land. They would meet it on the runway and retrieve the film from a mission. They would then look at thousands of photos with a magnifying glass. 'How times have changed! But back then, it would have been hard to imagine that we'd someday be able to take pictures with our phones!

Photo Credits - p. 1 USAF, p. 2 USAF, p. 3 CIA, p. 4 USAF, p. 5 USAF, p. 6 US ACE, p. 7 USAF, p. 8 USAF, p. 9 USAF p. 10 USAF p. 11 USAF, p. 12 USAF, p. 13 USAF, p. 14 Central Intelligence Agency, p. 15 USAF, p. 16 USAF, p. 17 USAF, p. 18 USAF, p. 19 USAF, p. 20 USAF, cover USAF, opposite page USAF, back cover USAF

The appearance of U.S. Department of Defense (DoD) visual information does not imply or constitute DoD endorsement.

U-2 Specifications

Length:	63 feet
Span:	105 feet
Height:	16 feet
Maximum Speed:	410 mph
Maximum Altitude:	70,000 feet

Engine:	One General Electric F118-101 Engine
Power:	17,000 lbf thrust
Maximum Take-off Weight:	40,000 lb
Crew:	one
Armament:	none
Range:	Over 7,000 miles
Fuel Capacity:	2,950 gallons

Planes That Changed History - Lockheed U-2

Sources

Lockheed U-2 Flight Manual, March 1, 1959, Approved for Release January 6, 2012

The CIA and the U-2 Program 1954 - 1974, Gregory W. Pedlow and Donald E. Welzenbach, History Staff Center for the Study of Intelligence, Central Intelligence Agency 1998

Debriefing of Francis Gary Powers February 13, 1962, Central Intelligence Agency, Sanitized and Approved for Release CIA RDP84B00459R000100020001-5

The Summit Conference of 1960: An Intelligence Officer's View, Sherman Kent, March 9, 2007 Central Intelligence Agency

United States Air Force, U-2 Specifications

www.ingramcontent.com/pod-product-compliance
Lightning Source LLC
Chambersburg PA
CBHW051831210526
45473CB00005B/1826